Wilma Lima Lira
Lucília Dias Pacobahyba

The use of thematic rooms or ambient rooms

Wilma Lima Lira

Lucília Dias Pacobahyba

The use of thematic rooms or ambient rooms

in public schools in Boa Vista, RR

ScienciaScripts

Imprint
Any brand names and product names mentioned in this book are subject to trademark, brand or patent protection and are trademarks or registered trademarks of their respective holders. The use of brand names, product names, common names, trade names, product descriptions etc. even without a particular marking in this work is in no way to be construed to mean that such names may be regarded as unrestricted in respect of trademark and brand protection legislation and could thus be used by anyone.

Cover image: www.ingimage.com

This book is a translation from the original published under ISBN 978-620-3-46754-3.

Publisher:
Sciencia Scripts
is a trademark of
Dodo Books Indian Ocean Ltd., member of the OmniScriptum S.R.L Publishing group
str. A.Russo 15, of. 61, Chisinau-2068, Republic of Moldova Europe
Printed at: see last page
ISBN: 978-620-3-65369-4

Dedication

To my roots, my mother Luciene, my grandparents Dilza and Miguel for the incentive, support, and the daily example of fight and faith in life.

THANKS

The realization of this work was only possible because during my undergraduate years I learned and experienced many things, the university became a second home for me, I couldn't have achieved any of this without the people who are part of my life and daily life. Here are my sincere thanks.

I would like to thank all the professors who made my trajectory real and made it possible for my dream to come true. I was lucky enough to be a student of some exceptional teachers and among these professionals I would like to highlight teachers Pablo Oscar Amézaga, Maria Aparecida Neves, Lucilia Dias Pacobahyba, Silvana Tulio Fortes, Fabiana Granja, and Frank James Araújo from the Biological Sciences course and teachers Carlos Menim and Lindivalda Feitosa from the Pedagogy course, for all their classes that added not only to my academic formation, but also to my life.

In special I thank two great professionals, my supervisor, Prof. Lucilia Pacobahyba, for guiding me so well and patiently for all the orientations, corrections, and contributions during the whole process of building my TCC and my co-supervisor, Prof. Maria Aparecida Neves, who since the beginning of the course has taught me a lot about what it is to be a teacher, warned me that it is a career where we face many difficulties, but that it is also very rewarding, when we perform with dedication and love, she will certainly always be a great example for me.

To Capes and the PIBID program for granting the scholarship for teaching incentive, because this action provides all undergraduate students an essential opportunity to have an idea of how it is in practice to be a teacher, it is a unique chance we have to be sure that we chose the right profession. In my case, this certainty is only reinforced by the fact that I never had any doubts about it, even with all the difficulties I faced during my undergraduate studies. In addition, we had the chance to meet some excellent teachers who work in basic education, such as Ilândia Perez Lima Cruz, who was my supervisor the entire time I was part of the PIBID.

To all the employees and servants, who contribute in the best way for the good functioning of this institution called Universidade Federal de Roraima.

I will not mention all of them, but there are some that I cannot fail to mention who started the course with me, such as Andressa Sampaio, Cleidiane Gomes, and Rodrigo Lopes, others are from later classes, such as Jairo Ferreira,

Ismitiely da Silva, Jaquelice Guerra, Jessica Martins, in short, everyone helped me a lot during this journey.

And finally, on a personal level my most important thanks. I thank my beloved family, who have always been by my side giving me support. Especially the strongest and hardest working woman I know, my mother, who is my example of strength, I will never be able to repay her for everything she did and does for me. Of course I could not fail to mention the two people I love the most in life, my daughters, they are the big reason I am here, because of them I went all the way to the end of the degree despite sometimes having thought of giving up, without them especially the eldest, Giseli who helped me a lot taking care of her younger sister,

Ana Vitória, whenever it was necessary so that I could come to college. Both of them were always by my side encouraging me in the realization of my dream, this conquest is not only mine but ours.

The ambient room should be a place to structure, organize, plan, and make thinking happen; it is a space to facilitate, for both the student and the teacher, to question, conjecture, search, experiment, analyze, and conclude, in short, to learn, and mainly to learn how to learn.

(SÉRGIO LORENZATO, 2009)

SUMMARY

This work aims to know the operation of the thematic rooms/environments of public schools in Boa Vista/RR that in the last 09 years have been implementing these differentiated teaching spaces, to verify how teachers are using such teaching spaces, a resource still little used in public schools in the Capital, besides knowing from the experience of the students if this model of organization is well accepted and if it brought improvements in the development of classes and consequently in the learning process. To this end, an exploratory field research was carried out. Five schools were selected, two elementary schools, two high schools, and one school that offers both levels of education, where structured questionnaires were applied to 65 elementary school students, 61 high school students, and 11 teachers. The results indicate that the implementation of environmental classrooms in Boa Vista schools is something that needs to be better studied, because despite being a new organizational model that is very well accepted and, according to the teachers who participated in the research, has a good physical structure, these spaces do not have the necessary resources and materials to perform their role. After all, it is necessary to combine infrastructure, material, pedagogical and social conditions. Only in this way is it possible to develop pedagogical practices that promote more interaction and exchange of experiences among the students and also between them and the teachers. The adoption of theme classrooms allows teachers to have more autonomy and change the school environment, which can bring about very positive changes not only as students but also as citizens.

Keywords: Environment Room, Teaching Resources, Proper Use

ASS Ayrton Senna da Silva
EF Elementary school
EM High School
LDA Lobo D'Almada
LDB Laws of Directives and Bases
MDA Mario David Andreazza
PB Penha Brazil
PCNs National Curriculum Parameters
PCS President Costa e Silva
RR Roraima
SC Santa Catarina

1. INTRODUCTION

The teaching practice is something that has always been studied over time, these studies have the purpose of promoting improvements in the teaching learning process. And recently, the thematic rooms have begun to emerge as an alternative that institutions can use as a tool to contribute to this process. These classrooms can greatly help the development of school practice, because it is a different way that provides teachers with the planning and development of differentiated, more elaborate, and interesting classes, as they can count on the materials available in these study spaces. The interest for this subject came up a while ago when during the General Didactics course I did a work where I talked about thematic classrooms, but in this case I only talked about the environment rooms of the School of Application, I found it very interesting how these study environments work and how they are in this school and when I thought of a theme to develop in my research project I thought it would be good to see how these environments work in public schools in our city.

Thus, it became relevant the study and knowledge of this new teaching strategy, which was the object of research of this work. To this end, a quali-quantitative and descriptive field research was carried out, through a questionnaire with teachers and students from public schools in Boa Vista/RR that are working with these teaching spaces.

What we did in this study was to research how these educational spaces are being used in public schools in Boa Vista-RR. Since this practice is recent in the public school system of our municipality. It was only in 2009 that this practice was implemented in the city. The first public school to adopt this form of teaching was the school of Application of the UFRR, which is a different case and is indeed a public school, but it is not a school under the responsibility of the State Government, it is an educational institution that is part of the Federal University of Roraima. In the same year the Lobo D'Almada school was the first among the state schools to implement ambient classrooms.

1.1 AND WHAT ARE THE THEMATIC ROOMS?

This is a thematic environment that can spark curiosity and interest in students that we don't see when it comes to a conventional classroom.

The idea of these ambient rooms is to organize classrooms with teaching materials pertaining to each subject, i.e., the school will have a room for Portuguese, mathematics, chemistry, science, among others. The objective of this organization of spaces is that each room be equipped with the resources and materials necessary to illustrate and enrich the classes, such as, for example, the sets of maps, photos and engravings in the geography rooms. The models, atlases of the human body, organ and animal models in the science room, microscopes, and so on.

The ambient classrooms are not limited as only a physical space, because all the materials and equipment available in these spaces are only the means for the teachers using these didactic and pedagogical resources to meet each one in their discipline in question, the specific educational purpose. The idea is to provide the student with a greater interaction with the diversity of educational resources and materials, and thus have the opportunity to establish a relationship between school knowledge, his life, and his daily life, the everyday life outside of school.

> A classroom in which didactic and pedagogical resources are available to serve a specific educational purpose. The idea is to make the student interact with a greater diversity of resources and teaching materials and be better able to establish a relationship between school knowledge, his life, and the world. Moreover, the Ambient Room concept considers that the "chalkboard or blackboard" is not the only valid resource in the teaching-learning process in the face-to-face way. (MENEZES; SANTOS, 2002 apud ALMEIDA 2012, p.6).

The ambient room is actually not something new in the school environment, according to teacher Sônia Penin (1997). This tool has long been part of the daily routine of some schools, but it is little or underused in many cases. It is very common to find specific rooms called laboratories in schools. For example, the computer lab, the science lab. However, in these spaces, classes are taught occasionally and not daily, as in the case of environmental classrooms.

The objective of this space organization is that each room, once specialized, each teacher has his or her own room, he or she can count on all the materials and resources necessary for the illustration and enrichment of the classes.

In this sense the classroom should facilitate an environment of continuous learning exchange for both students and teachers. Since these spaces can be used and changed according to the needs and goals of each class.

And when we talk about changes in teaching practices we cannot forget that teachers must be prepared, and for this the essential thing would be that they seek whenever possible to be improving themselves to know how to deal with these changes, such as the opportunity that arises when the institution adopts thematic classrooms. This improvement and preparation is something demanded by society, which seeks professionals trained to develop an integrated, updated, contextualized and interdisciplinary work, for a complete education where one does not just learn fragmented content.

The classroom is an environment that must be organized according to the teacher's needs, because if the teacher has this space and the necessary materials to teach the class, he or she will not waste time between class breaks.

With this in mind, school institutions are becoming more and more open to the idea of setting up ambient classrooms. It is necessary to take into consideration the stimulus of all areas of knowledge, because this will contemplate the students in a general way and will make the teaching-learning process differentiated and directed with alternatives of activities appropriate to the various areas of knowledge, besides giving them the freedom in choosing and handling the available teaching materials.

When the school organizes the school space in thematic classrooms, it ensures that all subjects can be better understood and assimilated by the students, thus making learning more meaningful.

Theme rooms are an excellent alternative for organizing these spaces, which can then try to provide students with differentiated, stimulating, and enjoyable classes, which can end up facilitating the interaction between students and teachers.

This study was justified, therefore, by the need to know if the public schools of Boa Vista/RR are managing to implement and make these educational spaces work according to the proposal of the thematic rooms

During the research I verified some important issues for a good functioning of the thematic classrooms, of course the most important was the main objective of the work that was to verify how is the physical and pedagogical structure of these school spaces.

Furthermore, I tried to diagnose with the teachers if the adoption of the ambient classrooms improved their praxis and also if it stimulated the students' participation and consequently their performance in class, considering that one of the most difficult tasks the teacher faces nowadays is to get the students' attention during the class.

For these environments to be considered thematic classrooms, they must have a structure with diversified materials corresponding to each school subject. The materials serve as subsidies for learning to take place, and the teacher is responsible for the mediation of knowledge.

Students, in their different ways of thinking and acting, need environments that involve and instigate them to learn, since nowadays the different media all the time remove students from human contact and distance them from interaction with classmates and teachers. As a future teacher, I know that we need to use technology to our advantage and, thus, thematic classrooms, when well organized, can help us a lot in this matter.

To develop the research we chose five state schools that adopted the ambient classrooms and applied a questionnaire to the science and biology teachers of these institutions and to some students from 7th to 9th grade of elementary school and from 1st to 3rd grade of high school. The objective was to verify how the schools that adopted this teaching methodology organized these study environments, both physically and pedagogically, with materials and equipment for the teacher to be able to develop a good job with their students, as well as to verify if the use of the thematic rooms has contributed to the teaching learning process, since this is its purpose.

2. REVIEW OF LITERATURE

According to Krasilchik (2000), science became a mandatory subject in the country after the Law of Directives and Bases of Education (LDB), No. 4.024/61. This law greatly expanded the participation of science in the school curriculum, which started to appear from the 1st year of junior high school. In high school, there was also a substantial increase in the number of hours of Physics, Chemistry and Biology. Only in 1971, with LDB No. 5. 692/71, did the natural sciences become compulsory in the eight grades of primary school.

The fundamental objective of science teaching became to enable the student to identify problems based on observations of a fact, to raise hypotheses, test them, refute them and abandon them when necessary, working to draw conclusions on their own. The student should be able to "rediscover" what science already knew, by appropriating its way of working, understood then as "the scientific method": a rigid sequence of pre-established steps. It was with this perspective that the democratization of scientific knowledge was sought, recognizing the importance of scientific experience not only for possible future scientists, but also for the common citizen (BRASIL, 1997, p 19).

> In the first grades of elementary school, each class has one teacher responsible for all subject areas. In the last four grades, biology is part of the science subject, which also includes the topics of physics and chemistry. In high school, the professional practice of biology in our country varied greatly in the decades from 1950 to 1990, biology was subdivided into botany, zoology and general biology, topics that composed with mineralogy, geology, petrography and paleontology the natural history discipline (KLASILCHIK, 2011, p. 12).

In 1996, the new LDB, No. 9.394/96, was approved, establishing in paragraph 2 of Article 1 that school education must be linked to the world of work and social practice. Article 26 establishes that "the curricula for primary and secondary education must have a common national base, to be complemented by the other curricular contents specified in this Law and in each educational system". The basic formation of a citizen in primary school requires full mastery of reading, writing and arithmetic, understanding of the material and social environment, the political system, technology, the arts and the values on which society is based.

High school has the function of consolidating knowledge and preparing for work and citizenship to continue learning (KRASILCHIK, 2000).

Besides the specific contents of the science areas, in 1998, the Secretary of Basic Education, through the National Curricular Parameters (PCNs) - Natural Sciences, presented four thematic axes that would guide the teaching of science at this level of education: Earth and Universe, Life and Environment, Human Being and Health, Technology and Society. In the same period, the PCNs - Transversal Themes was launched, proposing new themes to be included in the curriculum: Ethics, Cultural Plurality, Environment, Health, Sexual Orientation (BRASIL, 1998a; 1998b).

Education in our schools has been throughout history undergoing transformations that reflect the changes that have occurred in society, politically, economically and culturally. Each new government brings reforms that mainly affect primary and secondary education, and these changes range from the system to the physical structure offered in school buildings, and among these changes we have a good example to be followed, the adoption of thematic classrooms or ambient classrooms.

Traditional schools are not concerned with the physical space of the classroom. Most of them have spaces composed only of desks and blackboards. This makes the environment unstimulating for the students.

Adopting a conception different from the still dominant practice, Moreira (2007) teaches that the school environment should be configured as a space socially constructed by the interactions between students and teachers and of them with the other material and symbolic sources of the environment. It should, therefore, be organized with the purpose of promoting learning opportunities. And this learning takes place not only through the "transmission" of content, but also through the exchange of experiences, through dialogue between those involved in this process.

In view of the proposal of teaching as interaction, the environment room comes as a new school organization model different from traditional classrooms, because it is specifically directed to a subject, with emphasis on the arrangement of teaching materials, in order to offer greater interactivity among students, so that they can build knowledge linked to reality.

When we learn about these contemporary perspectives, we can research the origin of environmental classrooms, as a place of interaction of technologies and educational methodologies, in the history of Brazilian education. According to Nunes (2000, p. 52-7), the use of environmental classrooms as a differentiated pedagogical alternative comes from the 1960s, through the experimental classes, based on the principles of the New School and the multipurpose gymnasiums, originated from American schools. Still according to Carvalho (1999, p. 15) the expansion of schools working with environmental classrooms in Brazil occurred mainly from the 1980s, in schools from 5thto 8th grade of Elementary and High School.

According to Sanfelíce (1998), the classroom is a specific place destined to specific teaching-learning activities of specific knowledge, in different levels and complexity, through appropriate methodologies and that only has its peculiarity assured in the measure in which teachers and students guarantee the real execution of these objectives to which it is destined. The classroom, then, is not that inert physical space of the school institution, but a physical space made more dynamic by the pedagogical relationship.

And what would a thematic room or environment room be? It is nothing more than an appropriate space where the materials necessary for the performance of their function can be found, that is, a classroom where the teacher can count on the support of some teaching resources that will help him/her develop the contents of his/her subject.

The environment room brings the student closer to the science in question, because it opens a range of possibilities that until then could go unnoticed to them, because they were not exposed to the knowledge (PENIN, 1997). These environments can greatly help teachers to develop their classes, especially science and biology classes that, according to their content, offer several options of materials and differentiated classes.

In high school, the PCNs have the "double role of disseminating the principles of curriculum reform and guiding the teacher in the search for new approaches and methodologies" (BRASIL, 1999, p.13)

The ambient classrooms are a tool that can be used in science teaching to make the class more interesting and interactive, because compared to traditional classrooms, these environments provide various teaching materials that will arouse the curiosity and interest of students, thus providing a more significant learning experience.

> An environment establishes a climate that predisposes a person to feel certain sensations, as well as willingness and predisposition to manifest specific behaviors, different actions, different attitudes. Planning a knowledge environment that invites people to learn and to enjoy the search for new knowledge is the task of teaching professionals (PENIN, 1997, p.20).

According to Costella, (2014) it is not possible to take students everywhere and show them the phenomena, the real forms of the spaces and how they are organized, but it is not from this impossibility, that we will be satisfied with a knowledge made based on readings of answers. Students are essentially symbolic, from the visual stimuli and this symbology that the ambient room offers support, works as a synthesis for knowledge and opens doors for learning possibilities.

The thematic rooms are an excellent resource that offers teachers some benefits, as mentioned before, with the use of these spaces the teacher has more time to teach their classes, the students have the chance to use many didactic materials that will help in the understanding and fixation of the approached contents, if the environment is well structured the students can become more participative and creative, and all these benefits certainly help both the teachers and the students in this difficult process that is teaching-learning.

And making use of thematic rooms/environments in science teaching can be very profitable, because it is possible to use various teaching resources that will make the class more profitable and I believe more productive, for example, in the science or biology room we can use panels or models of human body organs, There are several possibilities and another thing to think about is that the way to organize the desks can be different instead of putting them in rows one behind the other we can use a circle.

At school, innovation includes not only curricular changes, but also the introduction of new teaching and learning processes, products, materials, ideas, and even people (HERNÁNDEZ et al., 2000, p. 29)

From the moment an institution decides to use ambient classrooms as a teaching methodology, it is necessary to check if the school has the physical structure, the financial resources and, mainly, it is extremely important to check if the teachers and other professionals in the institution are prepared to work with this new way of teaching.

The organization of school space in environmental classrooms does not guarantee a priori changes in the teaching and learning process. It is not restricted to the distribution of the physical space, nor to changes in the pedagogical practice developed in this context.

The changes need to be defined according to the needs and objectives of the moment and according to the nature of the subjects and the theoretical perspective followed by the teachers. It refers to a set of changes in both the physical and social dimensions of the school. According to Rosa (1997, p. 23) "when using the environmental room, we must problematize reality and build with students assumptions that will serve as guidelines for them to act as agents of change in the environment in which they live.

Planning an environment, whether it is in personal or social life, such as the school environment, requires knowing how to make the association of structure, space, and the expected function of this place, so that the objectives can be achieved and it is possible to develop the actions programmed for the environment. According to Penin (1997, p. 20) "the environments are carefully planned to invoke sensations and summon the actions that they actually provoke".

The implementation of workrooms requires discussion and planning with all sectors of the school. It requires an exchange of experiences among the educational coordinators, managers and teachers who will be responsible for organizing the space for the workrooms in their schools. If this discussion and everyone's involvement is not present, it will certainly be difficult to succeed in the implementation process.

Speaking again about the definition of the environment room, Penin (1997, p. 20-1) emphasizes the need to analyze the focus of two dimensions underlying the environment room: the physical dimension and the social dimension. In its physical dimension, the environment-classroom constitutes a specific way to organize the pedagogical work of the several curricular contents so that certain differentiations may occur in relation to the traditional routine. Such changes would be centered on the form. Thus, the room-environment may be considered a physical (spatial) organization different from the conventional model, as it allocates the necessary resources to the school work of each subject, so that the access to didactic and pedagogical resources is easily available to teachers and students. In this sense, the environment-room is different from the conventional classroom because it concentrates equipment, books, maps and other resources required for the specific work with objects of knowledge of the various subjects.

Penin (1997) values the social dimension of the environment-rooms as a foundation for the re-dimensioning of learning spaces. The social dimension involves the arrangement of people in the room and the type of interlocution that takes place there. In this sense, the environment room is understood as a space in which differentiated pedagogical practices may result from this unconventional arrangement, enabling the teacher to work with school knowledge and with the students according to a new technical arrangement. In the pedagogical activities different relationships can be emphasized. The room-environment conceived as a space that facilitates learning allows a multidimensional pedagogical relationship, according to Penin (1997, p. 21). This relationship takes place in the interaction of the teacher with the students, in the exchanges with other areas of knowledge, to avoid isolation among the other subjects, and especially in the interactions among colleagues. The room reveals a posture open to the establishment of relationships between subjects (teachers and students), providing a stimulating coexistence in a more cheerful and beautiful working environment, rich in visual stimuli and messages.

3. METHODOLOGICAL PROCEDURES

3.1 METHODOLOGY

This study was based on the principles of a quali-quantitative and exploratory field research - we know that this type of research does not focus on numerical representativeness, but rather on deepening the understanding of its object of study. For Minayo (2001) qualitative research works with the universe of meanings, motives, aspirations, beliefs, values and attitudes, which corresponds to a deeper space of relationships, processes and phenomena that cannot be reduced to the operationalization of variables.

The instrument used for data collection was a semi-structured questionnaire, composed of nine closed questions and one open question whose objective was to allow the subjects participating in the research to justify their yes or no answer to the question: Do you think that the structure of the classrooms offers favorable conditions to facilitate the teaching learning process. The students, on the other hand, were asked whether or not they liked studying in the thematic classrooms, and according to the choice of yes or no, they should justify their answer. The study involved students from 7th to 9th grade of elementary school and from 1st to 3rd grade of high school. The choice of this public was due to the fact that they are more mature to answer the proposed questions. The questionnaire was the same for high school and elementary school students.

To get to know the perceptions of students from the state school system who study in environmental classrooms about this form of organization of the school space, I used a questionnaire containing 10 questions, where 01 of these questions was open and thus allowed the obtaining of a justification, contribution or opinion from the subject/informant, in addition to the standard closed response, obtained in the other 09 questions. Likewise, the science and biology teachers of these schools also answered a questionnaire with the same structure, but with some different questions.

To compose the research universe, we identified the Basic Education schools in the city of Boa Vista, Roraima, belonging to the public school system whose structure is organized in environmental classrooms (called thematic classrooms). Thus, we delimited the sample in five schools, two of them being elementary schools (Escola Estadual Penha Brasil and Escola Estadual Presidente Costa e Silva), two high schools (Escola Estadual Ayrton Senna da Silva and Escola Estadual Lobo D'Almada) and one that offers both levels of education (Escola Estadual Mario David Andreazza).

In total, 126 students participated in the research, 65 in lower school and 61 in middle school, and only those students who presented the Informed Consent Form (TALE) duly signed by their parents or guardians were accepted as participants. The 11 teachers participating in the research also signed a Term of Free and Informed Consent (TALE). As a method of analysis, we followed the parameters of qualitative research, seeking to know the opinions of students and teachers who participated in the research regarding the implementation, organization, use and influence of these differentiated, the environment rooms in the teaching-learning process.

To facilitate identification, I will identify the schools that participated in the research as A (Penha Brasil), B (Presidente Costa e Silva), C (Ayrton Senna), D (Lobo d' Almada) and E (Mario David Andreazza).

1.2 CHARACTERIZATION OF THE SCHOOLS PARTICIPATING IN THE RESEARCH

The Penha Brasil State School (A), is located at Rua Juscelino Kubitschek, 926, Aparecida neighborhood, North zone was founded in the government of Colonel Hélio da Costa Campos and created by decree No. 030 on June 7, 1973. On November 1st, 2001, after a long reform the school was re-inaugurated. Today the level and modality that offers its public is only elementary school, the final series from 6th to 9th. The institution works with thematic classrooms since 2010.

School B (Presidente Costa e Silva) is located at Rua Professor Agnelo Bitencourt, 719, São Francisco, in the North Zone. The school implemented thematic classrooms 6 years ago.

The Ayrton Senna da Silva school (C) is located at Rua Floriano Peixoto, 221, Centro and was founded in 1994.

The (D) Lobo D'Almada school is located in the area at Avenida Benjamim Constant, 1453, Centro. The school was founded in the mid-1940s, more precisely in 1945. This historical fact is recorded in the Diário Oficial do Estado, published on April 20 of that year. At 9am that day the Grupo Escolar Lobo D' Almada was inaugurated, and its first teachers were Jacobede Oliveira, Maria C. Matos, Lindalva Liberato and Isnal Andrade. They started teaching classes to the first 129 students of the Territory. Currently, it only offers regular high school classes in the morning and afternoon shifts, with classes distributed in 15 classrooms. Since 2009, the institution has adopted the use of open classrooms.

The Mário David Andreazza State School (E) is located at Rua Alcides Lima, 246, Caimbé neighborhood, in the West Zone, was established by Decree 020/1988, and offers the community the final years of elementary school and regular high school. Currently, the school operates in two shifts, morning and afternoon, and has 12 classrooms. It implemented theme classrooms 5 years ago.

4. RESULTS AND DISCUSSIONS

I will start talking about the results of the study by talking about the teachers. The questionnaire began by asking them if when the school decided to implement the thematic classrooms, they had had any kind of training. According to the answers, 45% had been using the environment for 1 year, 10% had been using it between 2 and 3 years, and 45% had been using it for more than 3 years.

When questioned about the physical structure of the classrooms and whether they offered favorable conditions to facilitate the teaching-learning process, most 82% answered yes, while 18% said no. In this question, after answering yes or no, there was an open question for teachers to justify their answer to the previous question. In this question, after answering yes or no, there was an open question for the teachers to justify their answer to the previous question, to say why they answered yes or no. Their justifications were And their justifications were: It facilitates the organization of materials, it is possible to prepare something different, assemble and use the four consecutive times, without wasting time, and they also highlighted that this structure arouses the students' curiosity.

The next question was the same as the one asked to the students about what elements they consider essential for the functioning of an environmental room, but with some different answer alternatives and just like the students most of them considered all the alternatives necessary and remembering that in this question they could mark more than one alternative.

Table 1. Response to question 4 asked of the teachers.

4 - Among the alternatives below, which elements do you consider essential for the functioning of an ambient room?	
ALTERNATIVE	**QUANTITY**
The physical structure of the institution	01
The organization of the room	02
The materials/resources available	03
The interaction with students and teachers	02
The students' interaction with the materials available in the room	02
All previous alternatives	08

The next questions in the teachers' questionnaire were to find out if the adoption of this new methodology contributes to an increase in student motivation. 64% answered that it contributed a lot and 34% that it contributed little. As for the improvement in grades, the students' performance, 73% said that there was an improvement in all subjects, 18% that there was an improvement only in some subjects, and 09% that there was no improvement. Finally, I asked the teachers if they thought that this teaching method should be implemented in other public schools in the city. The majority 55% answered that it depends, if the schools have the physical structure and teaching materials for the implementation, I agree with the implementation.

To know the perception of students about the use of thematic rooms, I initially tried to find out if they like to study in an environment room, the answer to this question was quite positive, because 87% of respondents said yes. After this first question, where the alternatives were yes or no, there was a complement to the question where I asked them to justify, to say why they answered yes or no. Among the answers, the students highlighted that this is due to the fact that classes become more interesting, fun, and contribute a lot to the interaction between students and teachers and among the students themselves, besides the fact that studying in ambient classrooms becomes easier and this change helps to get out of the routine that exists in a normal classroom. Only 13% of the students stated that they don't like the classrooms.

1 -Do you like the classes in the theme rooms?

YES - 109   **NO - 17**

Only 17 students out of the 126 who participated in the research answered that they don't like to study in thematic classrooms, and they justified their answer by saying that they are not well structured, that they don't have the necessary materials and equipment, and that there is no organization when changing rooms. The following chart shows the comments contained in the students' answers to this questionnaire:

Table 2. Responses to the open-ended question asked of the students.

Comments contained in the students' answers regarding the question: Why do you like or dislike studying in the thematic room.	Quantity	Percentage %
It delays classes, is very tiring and lacks organization. Sometimes the subjects get out of focus and are not well explained.	2	1,6%
Because I have improved a lot and it is much more interesting that way. It has several positive points, it is better for us and for the teachers. It brings more benefits.	3	2,4%
The school has too many students, it would be easier if the teachers changed class, the corridors are too narrow, it takes longer to start a class. I don't like to move to the other room.	5	4%
Because yes, I can meet my friends in the hallways, at class change, get a drink of water, go to the bathroom, and chat a little with my classmates.	5	4%
It is to change the study environment, to get out a little when we change class, to walk around a little and not spend all our time in one room. To escape a little from the classroom.	8	6,3%
We don't have theme rooms, we don't use them. I don't even want to study.	8	6,3%
Because learning is more relaxed, it is fun, it gets out of the routine, it is easier to understand the contents.	9	7,1%

	12	9,5%
Because I like to learn new things and new subjects, it shows that with new methods we can broaden our knowledge, making subjects easier. Sometimes you just have to look at the walls to remember something. We learn more by making the association of knowledge with the didactic subjects. We become more motivated to study, which facilitates our learning.		
Because it is good to have a room with its own identification comparable to the subject studied, each subject has its own room and we can differentiate one room from another. It is a different way of teaching, the decoration changes the environment, the teachers have freedom to decorate and organize their room.	11	8,7%
Because we can interact more with classmates and teachers. There is more organization and materials, there is more creativity in these spaces. They are cool, they help me memorize the class objectives, I stay more focused, I get more involved.	18	14,3%
Because the classrooms are comfortable, prettier, different, more interesting and enjoyable.	19	15%
DID NOT ANSWER	26	20,6%
TOTAL	**126**	**100%**

Similar results were found by NAVES, (2014) in research entitled "Room environment for teaching in geography: a case study. Course completion work (monograph), geography course. Whose objective was to conduct an analysis of the implementation of the ambient rooms, especially the geography room for students in the final years of elementary education from a case study of the Batista pereira Municipal Elementary School, located in the city of Florianópolis, SC. It was also discussed the role of the school and the teacher nowadays and how the insertion of new didactic tools can contribute to geographic education. According to this work, the idea of implementing the environment rooms in this school arose from a proposal by one of the geography teachers at Batista Pereira about six years ago. However, only in 2013 it was carried out on an experimental basis and the result was so productive that the teachers of other subjects became interested in the idea and requested the environment room for all other subjects, which ended up being done at the beginning of 2014. According to the guidance counselor, "The adaptation period was better than expected, the students reacted well to the change of space. There was no turmoil during the transfer of students from one room to another during the adaptation period. In an evaluation carried out at the pre-collegiate meeting the response to the change was positive, the implementation of these environments for the majority was productive and this will contribute to the quality of the teaching and learning process.

As for the elements that students consider essential for the functioning of an environmental room, they could choose between five alternatives as shown in the chart below.

Table 3. Essential elements of an ambient room

2 - Among the alternatives below, which elements do you consider essential for the functioning of an ambient room?

	QUANTITY	
ALTERNATIVE	**E.F**	**E.M**
The organization of the room	26	14
The teacher's way of teaching	25	23
The materials that make up the classroom	08	07
Interaction with colleagues and teachers	16	11
The students' interaction with the materials available in the room	10	05
All previous alternatives	28	31
Total	**113**	**91**

With this initial understanding, the next questions were to know, through the students' perception, if the adoption of this new methodology has contributed to and favored the learning process. 62% of the students affirmed that the thematic room facilitates learning. For 16%, the adoption of this new teaching method did not make learning easier. The other 22% answered that it helped a little, but only in some subjects.

Finally, as far as the students are concerned, I asked them if they thought that the teachers were prepared to use this new teaching strategy. And once again the answer from the majority was positive, as 58% of the students who participated in the survey said yes.

Table 4. Students' answers to question 10

10 - Do you think the teachers are prepared to use this new teaching strategy?

	QUANTITY	
ALTERNATIVE	**E.F**	**E.M**
Yes, some teachers know exactly how to make use of the available materials and give great lessons	41	31
No, not everyone is ready	10	14
Lack of preparation for some teachers; some don't know or don't like to use all the available resources	11	10
None of the previous alternatives	03	06
TOTAL	**65**	**61**

The questionnaires also presented two questions in common for the two groups participating in the research (teachers and students). The first question asked them about the change of classrooms, whether at the time the class times are changed, this action is done in an organized manner. Most teachers (06) said yes, this always happens in an organized way. Of the 65 elementary school students, 38 answered that the organization at this time is good, while for 31 high school students there is no organization at the time of the class change, as shown in the following chart.

Table 5. Response to the 1st question that was asked to both groups.

4 - How is the organization of the students when it's time to change classrooms?			
QUANTITY			
ALTERNATIVE	TEACHERS	E. F.	E.M.
GOOD	05	38	19
BAD	00	10	11
It is always organized	06	04	00
There is no organization	00	13	31
Total	**11**	**65**	**61**

I also asked about the materials that these spaces have available in these ambient rooms and among the several alternatives that they could mark (posters, magazines, television, computer, blackboard, data show, paintings on the walls, anatomical models, games, maps, etc.) the most marked alternatives were blackboard, posters, and materials made by the students.

5 - Among the following options, which of the materials are present in the ambient rooms of your school?			
QUANTITY			
ALTERNATIVES	**TEACHERS**	**E. F.**	**E. M.**
BLACKBOARD/WHITEBOARD	11	57	56
CARTELS	08	51	50
TELEVISION	02	10	08
MAGAZINES	03	10	02
OMPUTER	01	17	12
MAPS	01	24	16
BANKS	01	09	01
DATE SHOW	04	23	31
WALL PAINTINGS	03	27	19
ANATOMICAL MODELS	03	03	04
GAMES	02	14	03
MATERIALS MADE BY THE STUDENTS	07	25	17
ATLAS OF THE HUMAN BODY	03	22	04
OTHERS	03	20	06

After conducting the research I could see that the ambient rooms of the institutions that participated in the study are rooms that do not have a good structure as regards the didactic part so in my view and according to some students as obtained in the questionnaires these environments are not considered thematic rooms, because most do not have any didactic-pedagogical material or equipment that should have in these spaces, in some there are closet, posters, some paintings on the walls and some materials made by students and nothing more. The following image (Fig. 1) shows the science room of the Presidente Costa e Silva School, where we can see that the only materials available are some posters of a work done by the students and a poster.

Fig.1. Science room of school B (PCS) Source: author 2018.

Fig.2 and Fig.3 are from the math room at Mario David Andreazza School, where we can see that apart from the curtains, the only different elements in the room are the painting on the wall with the name of the subject and the signs that are present in mathematics: addition, division, subtraction, multiplication, and equality, and some numbers and musical notes to show that mathematics is present in music.

Fig. 2. Mathematics room at school E (MDA) Source: author 2018.

Fig. 3. Mathematics room at school E (MDA) Source: author 2018.

In the following pictures (Fig.4 and Fig.5) we have images of the resources that are part of the geography room at Penha Brasil School, the walls are completely covered by posters of works done by the students related to the contents, besides some maps and although it doesn't appear in the picture the room also has a steel cabinet.

Fig.4 geography room of school A (PB) Source: author 2018.

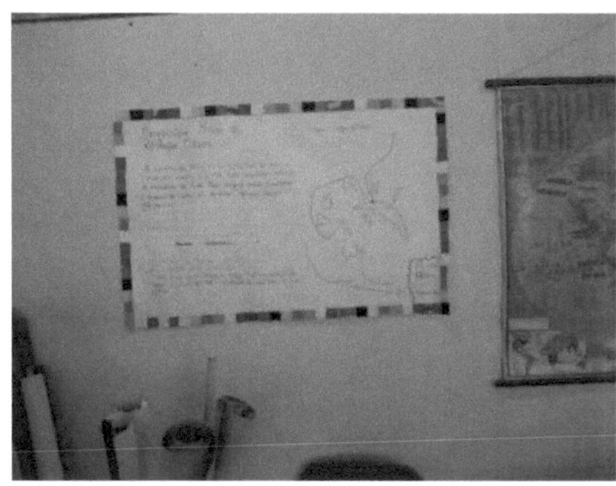

Fig.5. geography room of school A (PB) Source: author 2018

Figures 6 and 7 show the representative illustration with pictures painted on the walls of the biology room at Lobo D'Almada School, they are pictures of some contents studied in the subject. Just like in the other subject rooms I visited there is a steel cabinet in the corner of the room as well and nothing else besides that.

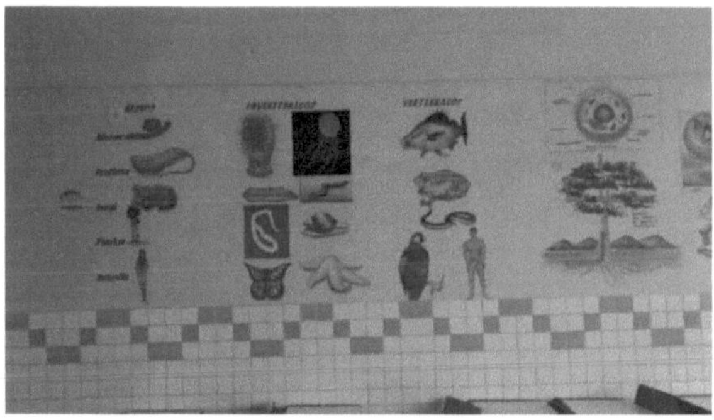

Fig.6 school D biology room (LDA) Source: author 2018.

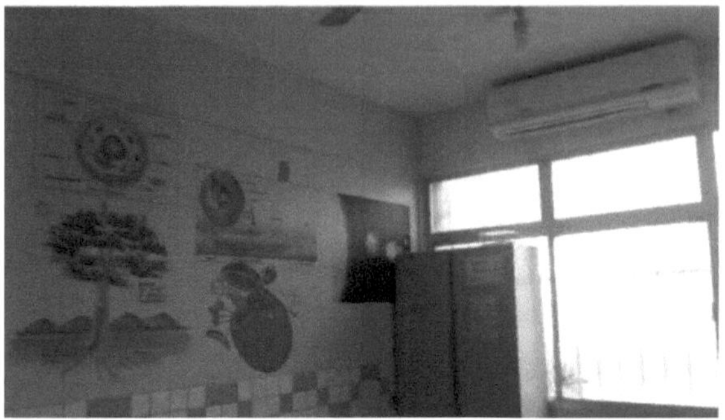

Fig.7. science room of school D (LDA) Source: author 2018

And finally, we have a picture that shows the environment of the Portuguese classroom at Ayrton Senna da Silva School. Among the five schools where I developed my research, I must say that this is the most precarious in terms of materials and resources, and I particularly wouldn't call these classrooms thematic, because they have nothing to make them fulfill the role of these environments (Fig.8).

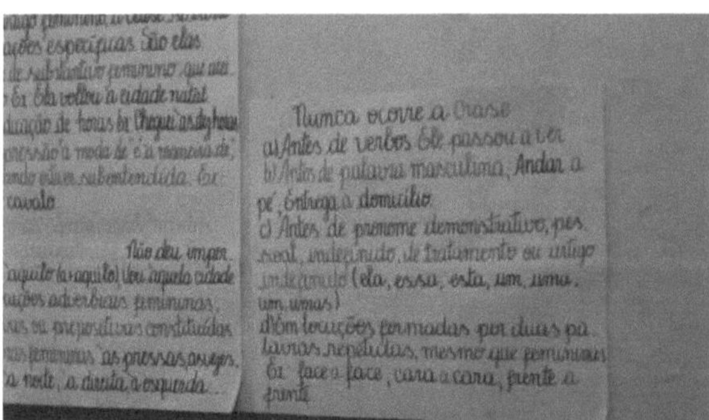

Fig.8 Portuguese room of school C (ASS) Source: author 2018.

To achieve the proposed goal for teaching in a thematic classroom, it is not enough for the school management to separate a room for each subject and say that when the class times are changed, the students will change rooms. In order for a classroom to function as an environment room, it is necessary that each room has materials and instruments that are appropriate to the subject taught in that classroom, and that can help and complement the development of learning in several ways.

According to CENP (SÃO PAULO, 1997) a Portuguese language classroom, for example, should have at the disposal of students a variety of reading materials: classic literature books, poems, newspapers, magazines, comic books, dictionaries, crossword puzzles as well as games structured and made by the students themselves that serve to contribute to the understanding and memorization of the diversities found in the mother tongue. Other indicated materials are puppets, costumes, props, and masks, in order to explore the students' narrative. With the use of physical resources, the ambient room can be transformed into a reading or text production workshop, in order to develop language activities (speaking, listening, writing, reading), activities of reflection on language (interpreting, constructing, organizing, structuring) and metalanguage activities (knowledge that builds grammatical theory).

If it is a specific biology classroom, it should be enriched with posters, models, charts, graphs, tables, anatomical models, skeletons, videos, photos, aquariums, terrariums, globes, microscopes, glassware, etc. Materials such as the human body charts, anatomical parts, the torso, and models of meiosis and mitosis, for example, are pedagogical resources that develop in the student the possibility of comparing different organs and systems, relating form and function, and establishing proportionality relationships. They are materials that bring the theoretical model closer to the real thing and their handling explores the sensory-motor, visual and aesthetic developments CENP (SÃO PAULO, 1997).

With regard to the factors that students consider important in a thematic classroom I could see that the result of my research was very much in line with what says (Rosário et al., 2014), according to their research 61% consider that all these factors are important in a classroom. In my research, besides the options: the teacher's way of teaching, the materials in the classroom, the interaction with classmates and teachers and all the previous alternatives, there was also the option of classroom organization, and as shown in the following graph, the option that had more votes was all the previous answers.

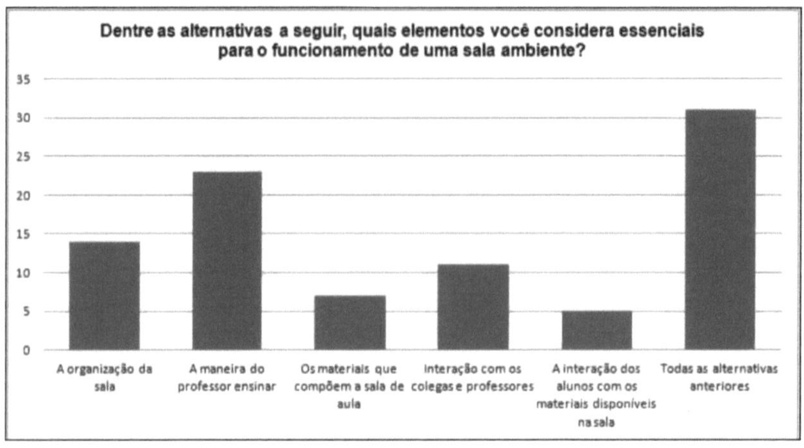

After understanding and comprehending what these educational spaces are and what the purpose of implementing them is, I see the big difference between what I should have found during the research and what I could actually see about the functioning of the ambient classrooms in the public schools where I developed my research. The right thing would be for me to see classrooms rich in didactic and pedagogical materials, with teachers having the opportunity to use different tools for the construction of student learning, thus developing creativity, performance, and interaction during the classes, making them more fun and enjoyable. But on the contrary, what I saw were mostly normal classrooms, and some still had some posters, a closet, and some other material made by the students, like some drawings on the wall, but nothing beyond that, and so I can't consider this environment as being a thematic room, because it doesn't meet the necessary requirements to be an environment room.

Regarding the teachers, according to what they answered in the questionnaire you can see that although they think it is necessary to improve a lot the didactic structure of these environments, they think that the implementation of ambient rooms has brought improvements in some aspects such as, for example, it is easier to develop the classes do not lose much time when they program something different and need to use some instrument such as the data show. As for the students, some teachers reported an increase in student interest, in some cases there was an improvement in their grades. For the teachers, if these classrooms are properly installed, this implementation makes a lot of difference, including improving their quality of life, since it is less stressful to teach in an ambient room due to all the amenities already mentioned.

After conducting the research it became clear to me, that although schools decide to adopt this new teaching strategy, decide to implement thematic rooms there is no study, a well worked out project that enables not only the implementation, but also the maintenance of these learning spaces. Talking to teachers and students I could see that at the beginning of the operation of these rooms some teaching and structural materials are made available, but unfortunately the school has no way to keep this environment tidy and organized as a thematic room should work. And so after some time this room no longer works with the purpose for which it was created since the only thing that remains of the initial organization is the fact that each teacher has at his or her disposal a classroom and at the time of changing times are the students to change rooms and not the teacher.

Another point that I think is important to highlight is the difficulty I had to do my research considering that most of the students are minors and, therefore, it is necessary that they bring signed by their parents a term authorizing their children to participate in the research. Soon after I went to the first school I realized that I would have a lot of difficulty and that I should, already thinking about this difficulty, deliver more terms than scheduled since the students took the terms home, but did not remember to bring them the next day. This becomes very clear when we observe the following graph that shows the difference between the amount of terms I delivered and the amount of terms I received back signed.

Moreover, I could also see that the students don't like to argue, to answer open-ended questions, because when asked to answer why they did or didn't like to study in themed classrooms out of the 126 students who answered the questionnaires, 26 didn't answer, they didn't justify why they did or didn't.

The results of the survey show that despite the advances, there is still much debate about the form and use of the ambient room. It is necessary that the school can discuss this option further. In this sense, this work also hopes to contribute to the debates that the school has been having to improve the service and the work with the students. The environment room alone cannot be the only option for change in a school, after all, it alone cannot meet all the needs found in the educational processes, which depend on other changes in the structure and infrastructure of the school, but also in the pedagogical and school management areas.

5. CONCLUDING REMARKS

The main objective of this research was to verify the functioning of these educational spaces in public schools in Boa Vista. It is important to study, know and introduce new teaching tools that stimulate students in their learning process, and this is not an easy thing to do. It is necessary that the pedagogical management and the teaching staff work together, develop a planning always through debates that respect the educational goals that should be the main focus and this planning must involve all sectors of the school.

The implementation of thematic rooms can make the teaching and learning process pleasant, fun, dynamic and enables the junction, the integration of content with practical experiences. When these spaces are adopted in the correct way, which unfortunately does not occur in the educational institutions that were part of this study. The conditions of the thematic classrooms in the five schools are very precarious both in terms of physical structure and in terms of the pedagogical aspect, in terms of materials and equipment available to teachers and students for a good development of their classes.

We cannot forget that these environments alone are not capable of making changes in the school's aspects, it is the students and teachers who are responsible for making these changes, and it is through them, especially the teachers, that it is possible to make these transformations.

After conducting the research it became clear to me, that these educational institutions that were part of this study can not make these classrooms environments work properly, according to the purpose of these spaces. In the beginning there is even an effort to make these environments work, but after some time the schools are not able to keep this good functioning, I believe this occurs because when planning the implementation there is no concern with the maintenance of these environments.

This reality could be changed if these teachers tried to overcome the lack of financial resources through the use of creativity, so that they could use these spaces exploring this creative universe, which offers several alternatives so that the teacher can organize the classroom in a way that accomplishes his or her goal, which is to enrich the learning process. In one of the schools where I studied, I saw a teacher who managed to do this; her classroom is very well organized and equipped with didactic and pedagogical materials that enable the development of her classes, her know-how process, her teaching practice. But unfortunately, only this teacher is managing to use these spaces properly; the vast majority of teachers end up letting the discouragement take over their teaching process and cannot take advantage of this new teaching tool.

REFERENCES

ALVES, F. M. G; COUCEIRO, K. C. U. S. **The importance of ambient rooms in teaching mathematics.** 2011. 16p. Curitiba.

ARRUDA, S. J.; CAETANO, M. R. Inovação curricular na escola pública:a teoria e a prática de Projeto Salas-ambiente. **Revista Universo Acadêmico, Taquara,** RS, v. 5, n. 1, jan./dez. 2012.

BRANDAO, C. R. A turma de trás. In: MORAIS, R. (org.). **Classroom - what space is this?** Campinas: Papirus, 1986.

BRASIL. Secretariat of Basic Education. **National curriculum parameters : natural sciences / Secretariat of Fundamental Education.** - Brasília :MEC/SEF, 1997.136p.

BRASIL. Secretariat of Basic Education. **National curriculum parameters**: history. Brasília, 1998.

CARVALHO, Leandro. **Room environment and the teaching of history.** Available at:< http://educador.brasilescola.com/estrategias-ensino/sala-ambiente-ensinohistoria. htm>. Accessed on: 29 July. 2018.

COSTELLA, R. Z. Competências e habilidades no contexto da sala de aula: ensaiando diálogos com a teoria piagetiana. **Cadernos de Aplicação, Porto Alegre**, v. 24, n. 1, jan./jun. 2014. Available at: < http://seer.ufrgs.br/index. php / CadernosdoAplicacao / article / view /23262/18279>. Accessed on: 12 feb. 2016.

GUERRA, V. P. **Pedagogical practices in high school: perspectives of teaching in environment-rooms.** Dissertation (Master in Education) - University of Tuiuti of Paraná Curitiba, 2007.

HALMENSCHLAGER, K. R. Thematic approach in science teaching: some possibilities. **Vivências: URI's Electronic Journal of Extension,** Vol.7, N.13: p.10-21, October/2011.

HERNÁNDEZ, Fernando et al. **Learning with innovations in schools.** Translation by Ernani Rosa. Porto Alegre (RS): Artes Médicas Sul, 2000.

LORENZATO, S. **O Laboratório de Ensino de Matemática na Formação de Professores.**

Editora Autores Associados Ltda. Campinas/SP, 2009.

KRASILCHIK, M. **Reforms and reality the case of science education.** 2000. 9p - Faculty of Education, University of São Paulo.

KRASILCHIK, M. **Prática de Ensino de Biologia.** 4. ed. ver. e ampliada, 3° reimpressão - São Paulo: Editora da Universidade de São Paulo, 2011.

MENEZES, E. T.; SANTOS, T. H. **"Sala ambiente" (entry). Dicionário Interativo da Educação Brasileira** - EducaBrasil. São Paulo: Midiamix Editora, 2002, Available at:< http://www.educabrasil.com.br/eb/dic/dicionario.asp?id=62> Accessed on: 29 Jun 2017.

MINAYO, M. C. S. (Org). **Social research: theory, method and creativity**. Petrópolis: Vozes, 2001

MOREIRA, A. F. **Ambientes de Aprendizagem no Ensino de Ciência e Tecnologia**. Belo Horizonte: CEFETMG, 2007.

MOREIRA, M. A. **Metodologias de Pesquisa em Ensino.** São Paulo: Editora Livraria da física, 2011.

MORAIS, R. D.(Org). **Sala de Aula: Que espaço é esse?** 22° ed.- Campinas, SP: Papirus, 1988.

NAVES, P. A. **Room environment for teaching in geography: A case study**. 2014. 59p. Trabalho de Conclusão de Curso(Graduation) - Universidade federal de santa Catarina, Florianópolis 2014.

NUNES, O. C. Ensino da história da educação e a produção de sentidos na sala de aula. **Revista Brasileira de história da educação,** n° 6 jul./dez. 2003.

PENIN, S. T. S. **A aula: espaço de conhecimento, lugar de cultura.** 3° ed - São Paulo: Papirus, 1997.

PENIN, S. T. S. Sala ambiente: invoking, summoning, provoking learning. **Revista Ciência Ensino.** Campinas, v. 3, p. 20-21, 1997.

ROSA, M. I. F. **Conversando sobre salas-ambiente no ensino de ciências. Revista Ciência e Ensino, Campinas,** FE/Unicamp, n° 3, p. 23-24, dez. 1997.

ROSARIO, C. L. et al. **Sala-ambiente: Espaço de Interação e Práticas Pedagógicas Inovadoras**. XI Symposium of Excellence in Management and Technology. Resende - RJ. 20014.

SANFELICE, L. J.; SAVANI, D.; LOMBARDI, C. J. **História da Educação: perspectivas para um intercâmbio internacional.** Campinas, SP: Associated Authors: HISTERDBR. 1998.

SÃO PAULO (State) Secretary of Education. Coordination of Studies and Pedagogical Norms. **A school with a new face: ambient-room.** São Paulo: SE/CENP, 1997.

APPENDIX A

Federal University of RoraimaCenter for
Biodiversity Studies-CBIOLcenciatura em Ciências
BiológicasDiscipline
: Research in Teaching of Sciences and BiologyAcademic
: Wilma Lima Lira

Questionnaire to be applied to public school teachers in the urban area of Boa Vista/RR.

Dear teacher, the questions in this questionnaire aim to collaborate with the implementation of the study: The Use of Thematic Rooms or Environment Rooms in Public Schools in Boa Vista/RR. This research is part of the Course Conclusion Work (TCC) of the Degree Course in Biological Sciences at the Federal University of Roraima (UFRR). The identity of the participant of this interview will be kept confidential.

Age: Sex: ( ) M ( ) F

Identification

School Name ____

1- When the institution decided to implement the ambient classrooms, was there any kind of training for the teachers?

( ) Yes, by the teachers' own initiative

( ) Yes, by initiative of the institution

( ) There was no training

2 - How long have you been using thematic rooms as a teaching tool?

( ) 1 Year ( ) 2 Years ( ) 3 Years ( ) More than 3 Years

3 - Do you think that the structure of the subject rooms offers favorable conditions to facilitate the teaching-learning process?

( ) yes   ( ) no

Why?____

4 - Among the following alternatives, which elements do you consider essential for the operation of a theme room/environment?

( ) The physical structure of the institution

( ) The organization of the room

( ) The available materials/resources

( ) The interaction with students and teachers

( ) The interaction of students with the materials available in the room

( ) All of the previous alternatives

5 - When it comes to changing class times, does this change occur in an organized manner?

( ) yes ( ) a little ( ) a lot ( ) no

6 - Among the following options, which of the materials are present in the room environment where you teach?

( ) Blackboard/whiteboard	( ) Data show
( ) Posters	( ) Painting on the walls
( ) Television	( ) Anatomical models
( ) Magazine	( )Games
( ) Computer ( )	( ) Materials made by the students
() Maps	( ) Atlas of the human body
( ) Benches	( ) Other

7 - Does the implementation of theme rooms contribute to the motivation of students for their studies?

( ) yes (   ) a little () a   lot ()   no

8 - Has this resource brought any change in the students' behavior during the lessons?

( ) Yes, the students are more attentive to the classes

( ) Yes, the vast majority showed more interest in the contents

( ) Yes, they now interact more in class and among themselves, which makes it much easier to approach the contents.

( ) I didn't notice any difference in the students' behavior

( ) None of the previous alternatives

9 -And regarding the learning process, are the students' results in the evaluations, after the implementation of the ambient classrooms, more positive?

( ) Yes, in all subjects

( ) Yes, but only in some subjects

( ) Yes, but the change was little

( ) Improved a lot

( ) There was no improvement

10 -In your opinion, should this method be implemented in the other schools in the City?

( ) Yes

( ) It depends, if the schools have the physical structure for the implementation, I agree with the implementation.

( ) No

( ) I prefer not to give an opinion

APPENDIX B

Federal University of RoraimaCenter for
Biodiversity Studies-CBIOLcenciatura em Ciências
BiológicasDiscipline
: Research in Teaching of Sciences and BiologyAcademic
:Wilma Lima Lira

Questionnaire to be applied to public school students in the urban area of Boa Vista/RR.

Dear student, the questions in this questionnaire aim to collaborate with the implementation of the study: The Use of Thematic Rooms or Environment Rooms in Public Schools in Boa Vista/RR. This research is part of the Course Conclusion Work (TCC) of the Degree in Biological Sciences of the Federal University of Roraima (UFRR). The identity of the participant of this interview will be kept confidential.

Age: Sex: ( ) M ( ) F

Identification

School Name ____

1 -Do you like the classes in the theme rooms?

( ) yes ( ) a little ( ) a   lot ( )no

Why?____

2   - Among the following alternatives, which elements do you consider essential for the functioning of an ambient room?

( ) The organization of the room

( ) The teacher's way of teaching

( ) The materials that make up the classroom

( ) Interaction with colleagues and teachers

( ) The interaction of students with the materials available in the room

( ) All of the previous alternatives

3 -Do you like this dynamic in which students change class in the intervals between

( ) Blackboard/whiteboard   ( ) Data show

( ) Posters   ( ) Painting on the walls

( ) Television   ( ) Anatomical models

( ) Magazine   ( )Games

( ) Computer   **classes?**

( ) Maps   ( ) Yes, because we have a little more time to drink water and go to the bathroom

( ) Yes, although all students at the same time in the narrow corridors sometimes leads to collisions between students and delays in class.

( ) Yes, because it is another opportunity to interact with colleagues

( ) No, I think it is a tiring and time-consuming process

( ) None of the previous alternatives

4 - Does the room exchange take place in an organized manner?

( ) yes ( ) a little ( ) a lot ( ) no

5 - Among the following, which of the materials are present in the ambient rooms of your school?

( ) Materials made by students ( ) Atlas of the human body

( ) Benches   ( ) Other

6 - In your opinion, did this teaching methodology make the classes more interesting?

( ) yes ( ) a little ( ) a lot ( )no

7 - Did this form of teaching promote a better understanding in your learning process?

( ) yes ( ) a little ( ) a lot ( ) no

8 - Did your grades improve after the implementation of the subject classrooms?

( ) Yes, in all subjects

( ) Yes, but only in some subjects

( ) Yes, but the change was little

( ) I improved my grades a lot

( ) There was no improvement

9 - Did this teaching strategy increase your motivation to study?

( ) Yes

( ) Certainly, because the classes became much more interesting and interactive.

( ) No, my motivation remains the same

( ) It increased a lot, because it is a differentiated and fun strategy

( ) None of the previous alternatives

10 - Do you think that teachers are prepared to use this new teaching strategy?

( ) Yes, some teachers know exactly how to make use of the available materials and teach great classes

( ) No, not everyone is prepared

( ) Lack of preparation for some teachers; some don't know or don't like to use all the available resources

( ) None of the previous alternatives

Printed by Books on Demand GmbH, Norderstedt / Germany